The Name Game

How Unethical Environmental Groups and Toxic Fanatics Scare You With Words

WRITTEN BY

Josh Bloom, Ph.D.

A publication of the

AMERICAN COUNCIL
ON SCIENCE AND HEALTH

American Council on Science and Health
110 East 42nd St, Suite 1300
New York, New York 10017-8532
Tel. (212) 362-7044 • Fax (212) 362-4919
URL: http://www.acsh.org • Email: acsh@acsh.org

Publisher name: American Council on Science and Health
Title: The Name Game: How Unethical Environmental Groups and Toxic Fanatics Scare You With Words
Author: Josh Bloom, Ph.D.
Subject (general): Science and Health
Publication Year: 2016
Binding Type (i.e. perfect (soft) or hardcover): Perfect
ISBN: 978-0-9910055-7-4

Acknowledgements

The American Council on Science and Health appreciates the contributions of the reviewers named below:

Hank Campbell
President
American Council on Science and Health

Martin J. Di Grandi, Ph.D.
Assistant Professor of Chemistry
Department of Natural Sciences
Fordham University

Jeremy Levin, Ph.D.
Medicinal Chemist (retired)

Steve Schow, Ph.D.
Vice President, Research and Development
Telik, Inc. (retired)

Christoph Zapf, Ph.D.
Senior Principal Scientist
Pfizer Global Research and Development

Table of Contents

The American public is perpetually being confused by terms, such as "organic," "natural," "man-made," "toxic," and "synthetic." Much of this confusion is both intentional and disingenuous—designed and promoted by unscrupulous environmental groups, giant companies that make dietary supplements and organic food companies, which have very successfully used fear-based, but scientifically baseless, techniques to advance their causes. Sometimes government agencies even get into the act.

Purpose of this book

This book identifies, exposes, and debunks the common tricks of wordsmanship that are used both to promote a self-serving agenda and to instill a misguided fear of chemicals in the mind of the American public. At the very least, readers should be able to identify these techniques, and be better equipped to evaluate the mis- and disinformation to which they are exposed.

1

Fear and ignorance: The Perfect Team

In 1990, only 49 percent of high school students[1] had ever taken a chemistry course. Twenty years later, that number had risen to 70 percent. The students from 1990 are now at the age where they are making decisions about their health and that of their families,' and some

1 "The Condition of Education." U.S. Department of Education, National Center for Education Statistics. (2015)

are even formulating science policy. Many of these decisions involve what food to eat, and which products to use.

It is all but certain that most of the 49 percent of these former chemistry students forgot everything they ever knew, probably not long after the final exam. The result is that very few adults in the US have even a marginal working knowledge of chemistry. When combined with other factors, such as the "organic," "natural," and "green" movements' efforts to further confuse matters, it is not at all surprising that we find ourselves in the middle of the "Perfect Storm of Fear," where people cannot understand what is safe to put in and on their bodies. Groups and individuals that benefit directly from these fears have been remarkably successful in propagating them, despite the fact that the fears are invariably, and often intentionally, based on flawed or nonexistent science.

Much of this phenomenon can be traced back to The Dietary Supplement Health and Education Act of 1994 (DSHEA)[2], in which Senators Orrin Hatch (R, UT) and Tom Harkin (D, IA) formulated a law that is as good an example of scientific doublespeak as you'll ever see. In order to support the enormous dietary supplement industry in Hatch's home state of Utah, DSHEA codified an artificial, and scientifically impossible, disconnect between drugs and supplements. The intentionally crafted and misleading language of this law resulted in product labels that were profoundly confusing and actually allowed manufacturers of these products to make false, sometimes conflicting, claims on the same label.

As the terms "natural," "organic," "drug-free," and "health aid" became omnipresent, so did the blurring of the lines between herbs, plants, drugs, chemicals and "non-chemicals," even though there is not (and can never be) any such thing as a "non-chemical."

2 Dietary supplement health and education act of 1994. Public law 103-417, 103rd congress https://ods.od.nih.gov/About/DSHEA_Wording.aspx#sec1

The nascent organic and "back to nature" movements contributed significantly to the increase in chemical ignorance and phobia. The soon-to-be colossal organic food industry, as well as opportunistic companies that manufacture every conceivable "all-natural green" product, had much at stake. For example, certain industries did a superb job of convincing the American public that it was somehow possible to lead a "chemical-free" life, by ridding oneself of products such as soaps, detergents, and disinfectants and replacing these with "green" products, like vinegar and baking soda, which not only have little utility as cleaners, but are both chemicals as well.

Fear Sells

The intent to instill fear of many perfectly innocuous chemicals, has two main, driving purposes: increase sales for companies that perpetuate the illusion that their products are somehow healthier because they lack some chemicals, and fundraise for groups whose existence depends at least partly on convincing the public that they will save us from the ubiquitous toxins that are relentlessly poisoning us.

An entire book could be dedicated to this effort. In fact, many have been written. A simple Google search for "fear monger books" yields more than 250 hits.

In recent times, examples of fear mongering are virtually limitless. There can hardly be a better example of a manufactured fear than what turned into the Y2K scare of 2000[3], when old software and code contained only two digits for the year, and were supposedly not equipped to handle the change from 1999 to 2000. Computers not reprogrammed would

3 "The Y2K scare: Causes, Costs and Cures." *Australian Journal Of Public Administration* · March 2004 https://www.researchgate.net/publication/24125256_The_Y2K_scare_Causes_Costs_and_Cures

think it was 1900. Companies knew this change was coming and were planning for it, but the media created a panic. Multiple catastrophic predictions, from the crashing of all computers to the collapse of the world power grid, provided almost unprecedented grist for doomsday prophets. *Newsweek* got a huge boost in sales with its 1997 front-page cover article "The Day the World Shuts Down[4]."

Contributing to the hysteria, The American Red Cross issued a preparedness checklist that, in retrospect, seems to be preposterous, but at the time contributed to the fear that gripped the nation. Among the dozens of supplies[5] that the agency recommended to have on hand were common sense items for any time, such as clean water, baby formula, insulin, and flashlights. The rest included "doomsday prepper" stuff: compasses, whistles, mess kits, syrup of ipecac (to induce vomiting in the case of poisoning), sleeping bags, pliers, and tents.

Sales of generators, wood-burning stoves, guns, and 55-gallon drums of water soared, but nothing illustrates the effect of fear on human behavior better than the following graph. Prior to Y2K, there was an astounding run-up in both the sales and price of gold coins[6].

As we ticked down toward the end of the century, the public fear was palpable. Yet, once the clock struck 12 and everything was just fine. Of course, there was an immediate collapse in the price and demand of gold coins, and a 55-gallon drum of water became a 456-pound paperweight.

4 The Day The World Shuts Down, *Newsweek*, June, 1997 http://www.newsweek.com/day-world-shuts-down-173474 Accessed 1/15/16

5 http://edition.cnn.com/1999/FOOD/news/12/09/food.prepardness/list.html Accessed 1/24/16

6 http://edition.cnn.com/1999/FOOD/news/12/09/food.prepardness/list.html Accessed 1/24/16

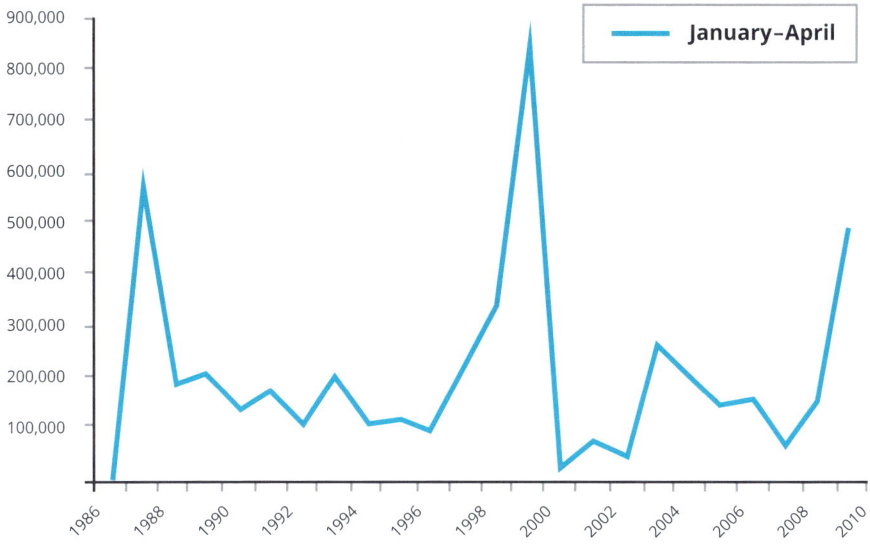

Gold American Eagle Bullion Coin Sale in Ounces
Source: Coin News

In the end, nothing happened, except that businesses that exploited fear for profit got rich. Fear sells.

A more recent, and especially repulsive, example of exploiting fear for profit involved a company that preyed upon parents' concerns during a time when there was a cluster of shootings at schools. Seeing an opportunity, the company began marketing bulletproof blankets[7] that were intended to be stored in student lockers. The very last thing that should

7 http://edition.cnn.com/1999/FOOD/news/12/09/food.prepardness/list.html Accessed 1/24/16

happen while an armed gunman is in a school is to have children running into the hallway for their lockers. In addition, data from the CDC[8] make it very clear that such events, although emotionally devastating, are very rare compared to real risks that students face every day. Student deaths from violence outside of school and automobile accidents are *far* more common, yet these do not elicit the emotional response of random shootings. This type of predatory tactic, which is clearly intended to focus attention on very small or hypothetical risks, can be very effective and also lucrative.

The "chemical scares" industry operates in a similar manner. It focuses on people's visceral fears of the unknown, and it would be difficult to find a better unknown fear than that of chemicals. Much of the public does not understand chemistry, and has been taught to fear it. Very few people are able to understand how their bodies dispose of the thousands of trace chemicals that are unavoidable in modern life. They may think they need to buy a "cleanse" to rid themselves of "toxins," unaware that human bodies do this just fine on their own. Taking advantage of the fear of the unknown enables groups with their own agendas to gain attention and funding—at the expense of actual evidence. Their actions further increase antipathy and distrust toward science, and especially chemistry. They use framing to undermine a company that "defends" a chemical that it manufactures or uses in a product, saying the company must be wrong, or biased, even though the defense is scientifically sound. When a chemical is "attacked" by an environmental group, bad, or even non-existent science, becomes "settled fact," and is then more than sufficient to create or perpetuate a scare. This is the essence of the chemical scare industry—selective use of evidence, which is often of poor quality.

One of the most disingenuous, albeit successful, tactics is the reason we wrote this book: how chemical fear mongers intentionally use chemical

8 "School-Associated Violent Death Study." Centers for Disease Control and Prevention, December, 2014 http://www.cdc.gov/violenceprevention/youthviolence/schoolviolence/savd.html Accessed 12/22/15

terms and names in a way that makes people draw incorrect and often scary conclusions. The goals of fear mongers are, unfortunately, made far easier by the confusion and ambiguity that are inherent in the nomenclature of chemicals. It is fair to call the system of naming chemicals "a mess."

Organic chemistry students in college are introduced to a variety of nomenclature systems—often quite different—that are commonly used to name chemicals. It is bad enough to have to remember that isopropyl alcohol is also called both "isopropanol," and "rubbing alcohol," but neither is technically correct; 2-hydroxypropane is correct. So is propan-2-ol. And, this is one of the simplest of organic chemicals in existence. When more complex molecules are named, things get exponentially more confusing. So, one can hardly blame the public for the failure to grasp an arcane system that can be perplexing even for trained chemists.

Unfortunately, this inherent complexity and ambiguity both provide ammunition for the chemical scares industry, and anti-science groups use this and other tricks very effectively.

Following are some of the common tricks used to scare us.

2

Intentional Conflation of the Name of a Chemical with the Chemical from which it was Made.

An old but remarkably effective tactic that is used by chemical scare-mongers is the essence of the "Name Game"— interchangeably, but erroneously, using the name of a chemical or product, and the chemical from which it was prepared.

By making use of this approach — a sleight-of-hand of chemical nomenclature — it becomes rather simple to manipulate the public into fearing a chemical substance based solely upon its name, rather than what really matters: the actual properties of the chemical itself. These scares typically arise when two different chemicals have similar names, and this makes it simple to demonize a harmless chemical that simply happens to share part of the name with an entirely different chemical that actually is dangerous. This strategy remains effective despite the fact that the two chemicals in question have nothing whatsoever in common— physically, chemically, or toxicologically—just part of a name.

A simple example of the "Name Game" can be illustrated with ordinary sodium chloride, table salt. Commercially, sodium chloride is obtained by mining. Although it would be absurd to do so, salt could also be made in a laboratory by combining sodium metal, and chlorine gas in the proper amounts. Sodium, a soft metal, which can be cut with a knife, is so dangerously reactive that it explodes when dropped into water. Chlorine—a green gas—was one of the chemical weapons used in World War I[9]. Both sodium and chlorine are very deadly chemicals, but when combined in the proper ratio will give 100 percent pure salt.

Now, imagine the natural foods industry trying to convince the general public that salt might spontaneously split into its constituent parts and explode on the dinner table, releasing a toxic gas. Although this is ludicrous, this same tactic is applied to other less commonly known chemicals. The resulting confusion can be easily exploited in a manner such that false scares are generated. This is the premise behind the use of manipulative, dishonest techniques that are calculated to suggest to consumers that toxic raw materials will necessarily produce toxic products, despite the fact that this notion violates basic principles of chemistry. The tactic also helps reinforce another important, yet equally false belief: that the salt that was made from sodium and chlorine is somehow different from salt that was mined or obtained from seawater. This is incorrect. They are identical in every way.

Polyvinyl chloride (PVC)

An often-used example of switching names is polyvinyl chloride (PVC), a widely used, durable and safe plastic, which is used in many common products, such as water pipes, construction, furniture, and clothing. PVC is manufactured from a toxic, carcinogenic gas called vinyl chloride. The similarity in the names is evident, but it ends there. PVC is not only *not*

9 How deadly was the poison gas of WW1? BBC News Magazine. http://www.bbc.com/news/magazine-31042472 Accessed 12/22/15

vinyl chloride, but the properties of the two are very different—an exceptionally stable white powder vs. a pungent, volatile gas.

When statements, such as "PVC is made from a carcinogen" are made, it is very likely that they are being done so intentionally. The statement "X is made from Y" should be a red flag that an individual, group, or agency is intending to put a black mark on "X" in the public mind, when, in reality, no such mark exists.

Styrene

Even more common is the interchangeable use of the chemical names "styrene" and "polystyrene." The former is a toxic, volatile, carcinogenic liquid, while the latter is the harmless plastic, which is also known as Styrofoam.

Three examples illustrate this obfuscation.

An article in the publication *Sustainable Business News*[10] entitled "Styrofoam Listed As Human Carcinogen" states the following:

"The US National Toxicology Program has decided to list styrene conservatively as reasonably anticipated to be a human carcinogen... We've long known that cancer rates are higher in and near styrene manufacturing plants."

This statement is technically correct, but is it also intentionally misleading? Yes, without question, since the headline states that Styrofoam itself is a carcinogen. It is not[11], but when combined with the added information

10 "Styrofoam Listed As Human Carcinogen" http://www.sustainablebusiness.com/index.cfm/go/news.display/id/25841 Accessed 12/22/15

11 Chemicals known to the state to cause cancer or reproductive toxicity, August 2015 http://oehha.ca.gov/prop65/prop65_list/files/P65single082515.pdf Accessed 1/15/16

that a higher incidence of cancer rates is found near styrene manufacturing sites, the take-home message for most people will be that styrene and polystyrene are one and the same, and are both toxic carcinogens, which is indisputably false.

The same article goes on to say: "[Styrene from polystyrene] can also leach into hot foods from cups, plates and clamshells." After seeing this, the average reader will not be able to make a clear distinction between styrene, polystyrene, and Styrofoam. All three will just be a jumble of cancer-causing chemicals. To determine whether there is any validity to the claims, direct or implied, that the use of Styrofoam containers exposes consumers to styrene itself, and that this puts consumers at risk, two questions must be answered: First, does Styrofoam contain residual styrene from the manufacturing process? Second, can Styrofoam degrade and revert back to styrene under conditions that would be encountered in real life?

Given advances in modern analytical technology, it is now possible to measure chemicals that are present in minuscule amounts, as low as parts per billion, and sometimes as low as parts per trillion. Common chemicals, which have been in the environment and in living organisms all along, still are there, although they could not previously be detected. Although it is indisputably wrong that the risk of a chemical has anything to do with the ability to detect it, antichemical groups routinely ignore this fact and wrongly equate the presence of a chemical with harm. The case with styrene is no different. Measurable traces of styrene can indeed be found in Styrofoam; however, the amount present is so small that its presence is completely irrelevant. The FDA[12] has discussed this in great detail:

12 "Safety of Polystyrene Foodservice Packaging." Food and Drug Administration: https://plasticfoodservicefacts.com/main/Safety/Safety-of-PS-Foodservice-Products/ Accessed 1/15/16

"Tiny amounts of styrene may remain in polystyrene following manufacture, so the FDA has evaluated both the safety of the food contact material itself (polystyrene) and the safety of the substance that may migrate (styrene). The result of these evaluations: FDA for decades has determined that polystyrene is safe for use in contact with food. In addition, the FDA has approved styrene as a food additive – it can be added in small amounts to baked goods, frozen dairy products, candy, gelatins, puddings and other food. [C]urrent exposures to styrene from the use of polystyrene food contact products remain extremely low, with the estimated daily intake calculated at 6.6 micrograms per person per day. This is more than 10,000 times below the safety limit set by FDA."

In other words, yes—there are miniscule traces of styrene that are released from Styrofoam, but in quantities that are far lower than those that could even conceivably cause harm. If further proof is needed, styrene occurs naturally in trace amounts in a variety of common foods[13], including beer, meat, nuts, and certain fruits. Once polystyrene is in use, it can indeed decompose and revert back to the styrene from which it was made. But this process begins at 625 degrees Fahrenheit[14]—not terribly different from the temperature on Venus (864 degrees). Regardless of your views on climate change, this is certainly not happening on Earth, so this concern can be dismissed outright.

The Earth Resource Foundation takes it one step further. Its publication entitled "Polystyrene Foam Report"[15] states: "Styrene is used extensively in the manufacture of plastics, rubber, and resins. About 90,000 workers,

13 "The Safety of Styrene in Selected Foods" https://plasticfoodservicefacts.com/Safety-of-Styrene Accessed 1/15/16

14 "The Thermal Depolymerization Of Polystyrene. Part 2—Formation of "Weak Links"" N. Grassie and W. W. Kerr, *Trans. Faraday Soc.*, 1959, 55,1050-1055.

15 "Polystyrene Foam Report" http://www.earthresource.org/campaigns/capp/capp-styrofoam.html Accessed 12/22/15

including those who make boats, tubs and showers, are potentially exposed to styrene. Acute health effects are generally irritation of the skin, eyes, and upper respiratory tract, and gastrointestinal effects. Chronic exposure affects the central nervous system showing symptoms such as depression, headache, fatigue, and weakness, and can cause minor effects on kidney function and blood. Styrene is classified as a possible human carcinogen by the EPA and by the International Agency for Research on Cancer (IARC)."

First, the entire paragraph above discusses styrene, not polystyrene. While the risks that are described hold true for industrial workers who regularly handle the chemical, they do not apply to coffee drinkers. This statement is nothing but a red herring.

Even the self-proclaimed miracle healer Dr. Andrew Weil chimes in[16]: "Polystyrene foam cups contain styrene — a chemical compound that is increasingly suspect. In the 12th edition of its "Report on Carcinogens," the National Toxicology Program (NTP) stated that styrene is "reasonably anticipated to be a carcinogen." Once again, this is nothing but a manipulative and amateurish scare tactic that equates the exposure of industrial workers to that of consumers. Weil's attempt to suggest that there is a relationship to takeout food containers is nothing but a big lie.

The intent of those who use this trick is perfectly obvious, yet the trick itself is nothing more than a bit of clever wordsmanship that is intended to foment a chemical scare when no such scare is warranted. Unfortunately, the trick works very well.

Even though we are scientists and not environmentalists, we note that there are *legitimate* environmental concerns about the widespread use of Styrofoam.

16 Andrew Weil, "Q and A Library" http://www.drweil.com/drw/u/QAA401095/Is-Styrofoam-Safe.html Accessed 12/22/15

Styrofoam is "infinitely stable." It does not chemically degrade in the environment, so it will never go away. Since the material is 95 percent air by weight, it is also very light, and this makes collecting and recycling the plastic especially difficult. Although it does not break down chemically, it does so physically, forming small pieces that will inevitably find their way into the ocean, lakes, and rivers. And, since it doesn't decompose, every Styrofoam cup that is manufactured adds to the amount of the material present on earth—an environmental one-way street.

Given these reasons, many people who care about a real environmental problem will avoid Styrofoam, and if the "Styrofoam haters" were concerned solely about the environment, they would not resort to scare tactics about styrene. This obfuscation undermines their credibility and should make people pause and wonder what their purpose really is.

"Antifreeze"

Another very effective example of using the "Name Game" to frighten people is antifreeze. There are two that are commonly used: ethylene glycol, which is toxic, and propylene glycol, which is anything but.

However, the Natural Resources Defense Council (NRDC) either doesn't know this or (more likely) does know it, and sees it as a fundraising opportunity using a cheap scare[17]:

"Engine coolants and antifreeze containing ethylene glycol and propylene glycol can be toxic and contribute high [biochemical oxygen demand] to receiving waters."

This statement is constructed in such a way that virtually anyone who reads it would conclude that both kinds of anti-freeze are the same, and both are poisons. This could not be further from the truth.

A fatal dose of ethylene glycol in humans is about three ounces (six tablespoons). Although, compared to many chemicals, this is only moderately toxic, there is an additional risk in this particular case: ethylene glycol is sweet (it is the simplest sugar), so animals and young children who taste it are likely to ingest more, thus increasing the likelihood of swallowing a dangerous or lethal dose. The reason that ethylene glycol is toxic is because it is converted by the body into oxalic acid, which then becomes highly insoluble crystals of calcium oxalate, which destroy kidney tissue. Calcium oxalate is the most common cause of kidney stones.

The story for propylene glycol is quite different. It is so safe that it is difficult to even find a toxic dose in humans. The CDC makes this abundantly clear:[18]

"Although propylene glycol is nontoxic under normal conditions, it can cause poisoning in rare and unusual circumstances. In one case,

17 "The causes of Urban Storm Water Pollution" http://www.nrdc.org/water/pollution/storm/chap2.asp

18 "Ethylene Glycol and Propylene Glycol Toxicity. What is Propylene Glycol?" http://www.atsdr.cdc.gov/csem/csem.asp?csem=12&po=14 Accessed 12/22/15

an 8-month-old infant with large surface area second-degree and third-degree burns was treated for many days with topical silver sulfadiazine containing a large amount of propylene glycol. The infant developed acute metabolic acidosis and cardiorespiratory arrest."

An 8-month old infant who has sustained second- and third-degree burns over a significant portion of her body has already suffered extreme trauma and is at great risk for infection and multiple organ failure. This case tells us nothing about the toxicity, or lack thereof, of propylene glycol, especially since it was applied topically rather than ingested.

But these do:

- "No studies were located regarding death in humans after oral exposure to propylene glycol."
- "A fatal case of propylene glycol poisoning occurred in a horse given 3.8 L of propylene glycol (one gallon) instead of mineral oil. The horse died of respiratory arrest 28 hours after administration."

If this farfetched safety profile does not sound "kosher" to you, think again, because it is exactly that. Propylene glycol is literally made kosher[19]; it is used in food. This is because, in a sense, it is food. Propylene glycol has such low toxicity because it is metabolized in the body to both pyruvic acid and lactic acid, which play a major part in biochemical reactions in all species of animals. Pyruvic acid is a key component of the Krebs (citric acid) cycle, which is the primary source of cellular energy. The Krebs cycle converts fats, carbohydrates, and protein into adenosine triphosphate (ATP) — the fuel that makes life possible. Lactic acid is a normal component of human metabolism, especially during times of energy expenditure. It is also found in many common foods.

19 http://www.sigmaaldrich.com/catalog/product/aldrich/w294004?lang=en&region=US

There cannot be a better example of obfuscation than intentionally equating two chemicals that happen to share a function (antifreeze), and part of a name (glycol), but otherwise could not be more different.

Intentional misuse of the word "chemical" solely to raise false scares.

Despite the old bromide "everything is a chemical," there is a wide-spread, vague belief that substances that are derived from nature are either not really chemicals after all, or a "different" type of chemical, which doesn't really count.

The use of the word "chemical" in a pejorative sense is exceedingly common, and is a very effective method of fear mongering. Unscrupulous environmental groups, peddlers of dietary supplements, and proponents of back-to-nature movements generously throw the term around.

Vani Hari ("The Food Babe") may be the worst offender to get public attention recently[20]: "There is just no acceptable level of any chemical to ingest, ever," she claims.

But she is hardly alone.

The Environmental Working Group (EWG)[21]: "Toxic iPhone: Apple's iPhone contains chemicals that are internationally regulated because of their potential danger to health and the environment, according to research by Greenpeace."

20 "The Food Babe: Enemy of Chemicals. How One Woman Mobilized An Army Against Food Additives, GMOs, And All Else Not "Natural."" Atlantic Magazine 2/11/15 http://www.theatlantic.com/health/archive/2015/02/the-food-babe-enemy-of-chemicals/385301/

21 IToxic? http://www.ewg.org/enviroblog/2007/10/itoxic Accessed 12/22/15

Mike Adams ("The Health Ranger")[22]: "Our bodies and cellular system are constantly under attack from a barrage of chemicals from the water we drink, the food we eat, the medical and dental industry, nuclear waste, etc."

Website supplement peddler Joe Mercola[23]: "More than 80,000 synthetic chemicals surround you every day. Synthetic chemicals are in your air, food, water, and in most of the products you use — many going straight into your body or your child's body, even before birth."

NRDC[24]: "As long as FDA doesn't know about the safety of thousands of chemicals, it cannot ensure the safety of the food we eat and cannot protect public health."

And, again from the NRDC[25]: "Chemicals are added to food to preserve, flavor, thicken, or otherwise alter it in some desirable way. Used this way, the chemicals are called 'food additives.' In other contexts, the same chemicals may be considered industrial chemicals or pesticides."

If ubiquitous chemicals were really poisoning us, we would be dying younger, and seeing increases in the incidence of cancer and other chronic diseases that would negatively impact our life spans. But,

22 NaturalNews.com "Healing And True Weight Loss Through Cellular Detoxification." http://blogs.naturalnews.com/healing-and-true-weight-loss-through-cellular-detoxification/ Accessed 12/22/15

23 http://articles.mercola.com/sites/articles/archive/2015/08/08/toxic-sludge-chemicals.aspx Accessed 12/22/15

24 "If You Think FDA Knows What Chemicals Are In Our Food, Think Again." http://switchboard.nrdc.org/blogs/mmaffini/chemicals_added_to_food.html Accessed 12/22/15

25 https://www.nrdc.org/experts/jennifer-sass/usfda-allows-chemicals-food-despite-lack-toxicity-testing

neither is true. Cancer rates have been steady or slowly declining for two decades, and we are living longer[26].

Use of the term "petroleum-based" to imply inherent toxicity

Another common, disingenuous scare term is the use of the word "petroleum-based." It conjures up images associated with the Exxon Valdez or the British Petroleum Gulf of Mexico oil spills, as if the crude oil is somehow part of the product from which it was made. It can be reasonably assumed that some consumers think that there is actual petroleum in a petroleum-based product.

This is not only false, but also reinforces the erroneous public perception that the origin of a chemical—natural or synthetic—has any bearing on the properties or health risk of that chemical. The correct definition of "petroleum-based" is quite simple: A chemical that is either isolated or purified from crude oil, or one that is synthesized from one of these chemicals.

A good example of a common petroleum-based chemical is aspirin. Aspirin does not occur naturally. Rather, it is synthesized from another chemical called "salicylic acid" using a simple one-step chemical reaction.

Salicylic acid (SA)—which itself is used for medical purposes, such as controlling dandruff and wart removal—may be obtained from the bark of a willow tree, or manufactured from phenol—a petroleum-based chemical that is synthesized from benzene and propylene—two of the many petrochemicals that make up crude oil.

26 "How Long Do You Have Left to Live at Age 65?" http://politicalcalculations.blogspot.com/2013/01/how-long-do-you-have-left-to-live-at.html - .VqKHp1MwjSc Accessed 12/22/15

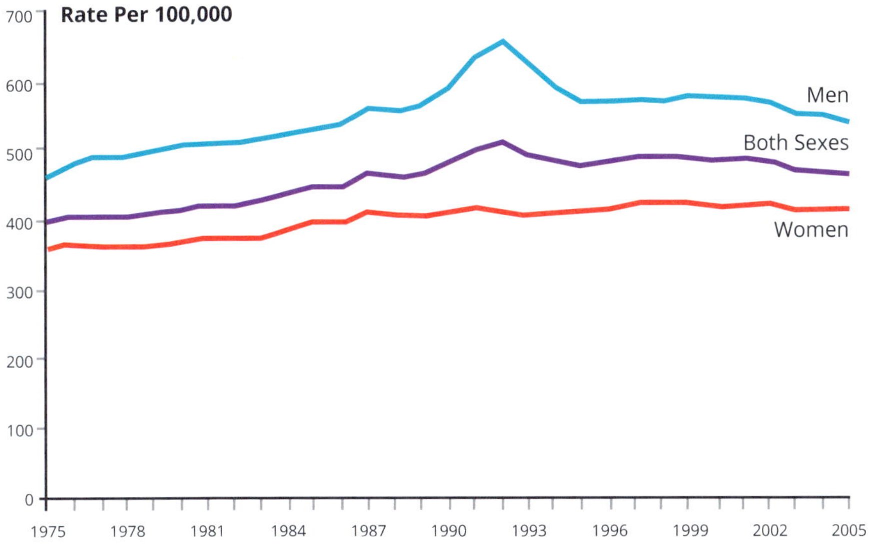

Cancer Incidence Rates* by Gender, US, 1975–2005

*Age-adjusted to the 2000 US standard population and adjusted for delays in reporting.
Source: Surveillance, Epidemiology, and End Results Program, Delay-adjusted Incidence database: SEER incidence Delay-adjusted Rates, 9 Registries, 1975–2005. National Cancer Institute, 2008

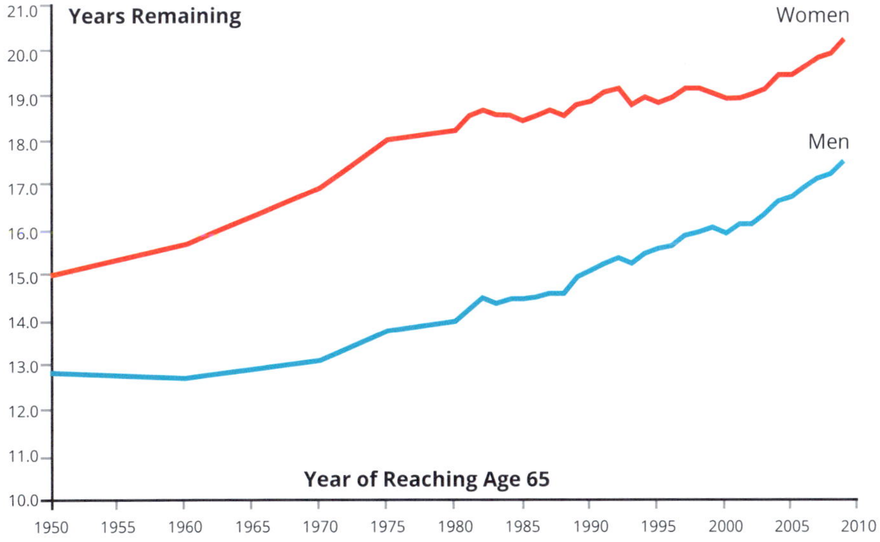

Remaining Life Expectancy at Age 65, 1950–2009

Source: CDC

The SA that is obtained from willow bark is identical in every way to the SA that is made from the two petrochemicals. They are indistinguishable.

> (Note: Chemicals containing carbon that are derived from living sources are infinitesimally more radioactive than the same chemical derived from oil. For the purposes of this book, this difference is irrelevant.)

The *only* relevance of the term "petroleum-based" in this (or any other) case is determining the source of the raw material used in the manufacturing process. In this case, petroleum wins hands down. There are not enough willow trees in the world to produce enough SA to make the 120 billion aspirin tablets[27] that are consumed each year.

Yet, this simple principle is either not understood or, more likely, fully understood and intentionally exploited by the Environmental Working Group, as evidenced by its fear mongering "Back to School Guide." Some examples include:

1) "Common crayons often contain paraffin wax, which is made from crude oil."

The group should be ashamed of itself. Yes, paraffin wax is made from crude oil, but this statement is not only totally irrelevant with regard to health, it is also blatantly manipulative. It would be very difficult to find a safer substance than wax. It is used in many foods, candy, chewing gum, and as a coating for cheese. It is not intended to be swallowed, but it doesn't matter if it is. The wax passes through the gastrointestinal tract and is excreted unchanged. Does anyone honestly believe that crayons are harmful?

27 Cyclooxygenase-3 (COX-3): Filling in the gaps toward a COX continuum?" Warner, T.D., Mitchell, J.A. PNAS(2002), 99(21), 13371-13373.

2) "Try to minimize kids' exposures to extra-strong or instant adhesives like epoxies, model and "super" glues; they contain toxic solvents. Water-based glues are safer bets, though most are made from petrochemicals."

Once again, the phony distinction of petroleum-based vs. natural is made, but at least EWG acknowledges that glue that is made from petroleum-based chemicals is safer, while contradicting itself in the process.

It is little wonder that parents may be terrified from reading propaganda like this. The number of examples in which this trickery is used is impossible to determine, but there is no doubt that it is a very large number. Why? Because it works.

Creating an artificial line between "natural" and "synthetic."

Drawing an arbitrary and artificial line between substances that just happen to be biosynthesized in a plant, microbe or animal rather than made in a lab is an extremely common scare tactic that is very effective, but scientifically flawed on every level.

"Synthetic" is portrayed as inherently unhealthy, implying it is not something your body is equipped to cope with. Conversely, "natural" strongly implies the opposite. Since we are supposedly "one with nature," natural must equal beneficial, despite that fact that many of the most toxic poisons on earth come from natural sources.

The Whole Foods grocery chain has this word game down to an art, banning artificial flavors, even though these are often the exact same chemicals that exist in the fruit or vegetable. But, a statement from the company's website[28], states: "Artificial flavors are produced through a set of complex chemical processes. They're intended to have the same taste and odor sensations as natural products, but chemically they are very different. No artificial flavors are allowed in the foods we sell. They just aren't necessary."

Whole Foods is either outright lying or ignorant when stating, "chemically they are very different." This is absolutely false, yet the company has no qualms about perpetuating bad science (and implicit scares) in order to carve out a niche in the highly competitive food industry.

Whole Foods will not take the *exact* same chemical that is responsible for part, or all, of the entire flavor of a food and add it to anything else. This is a logical and scientific abomination.

Perhaps most telling is that the company will not sell a product to which isoamyl acetate—the principal flavor of banana—has been added, even though isoamyl acetate is the same chemical that is found in the banana. It may or may not have even been derived from actual bananas, but this is conveniently omitted. Why? Because artificial flavors are "bad," even though most are also components of natural flavors.

The only real distinctions between flavoring a product with one of the chemicals that is identical to one of the flavors in the fruit, regardless how it was obtained, are flavor and scent. This is because, like wine, the taste and scent are dependent on a complex mixture of chemicals that give it its distinct properties. A pastry flavored with isoamyl acetate (banana oil) alone will not taste exactly the same as a banana. Sure, you will think "banana" when you put it in your mouth, but it won't taste the same, since many of the other chemicals present in a banana will be absent. Thus, the real issue is not the safety of artificial, or added flavors, but, rather, the flavor.

28 http://www.wholefoodsmarket.com/blog/our-quality-standards-no-artificial-colors-or-flavors Accessed 12/22/15

Ironically, bananas contain quite a few more chemicals, some of which are listed on the next page[29]. In what borders on surreal, some of these same chemicals in the real banana are themselves fodder for phony chemical scares.

Since it is unlikely that many of us fear bananas, it is interesting to examine some of the chemicals in them, and how they have been demonized. The table below identifies health scares that are associated with chemicals that are found in bananas. The discrepancy between the imaginary health scares and the obvious real ones—non-existent—is eye opening, yet it is precisely this type of phony scare that makes such an effective marketing tool. Chemical scares equal increased sales of organic products.

3

Chemicals: The dose does not make the poison. Except that it does.

Of all the chemical scare tactics that are used, this may be the most effective. It is also an example of the poorest science that is used by scare groups. This ploy is blatantly manipulative. It is based on "science" that is so flawed that it borders on voodoo.

29 http://blog.fooducate.com/2014/01/31/look-at-all-those-chemicals-in-a-banana/ Accessed 12/14/15

Selected banana chemicals with "bad reputations"

Banana Chemical	Function	Scare or complaint	Scare-monger	Reference
Fructose	A type of sugar	Cancer, liver damage, kidney stones, inflammation, impaired memory, gout, hypotension	PaleoLeap	http://paleoleap.com/10-reasons-why-fructose-is-bad/
Maltose	A type of sugar	Turns into fat	Bill Misner	http://tinyurl.com/ntglzvf
Glutamic acid (aka MSG)	Amino acid that is required for all mammalian proteins	"MSG: Is This Silent Killer Lurking in Your Kitchen Cabinets?" Headaches, neurotoxicity	Joe Mercola	http://tinyurl.com/6mzsj49
Aspartic acid	Amino acid that is required for all mammalian proteins	Endotoxicity	Joe Mercola	http://tinyurl.com/3z9b5le
Phenylalanine	Amino acid that is required for all mammalian proteins	Mental retardation, brain seizures, sleep disorders and anxiety	Vani Hari (The Food Babe	http://tinyurl.com/pgvqwtq
Stearic acid	Fatty acid	Skin irritation, organ toxicity, neurotoxicity, respiratory system problems	SteadyHealth	http://ic.steadyhealth.com/stearic-acid-side-effects
Myristic acid	Saturated fatty acid	Inflammation, elevated lipids, atherogenesis and vascular disease	David Katz, M.D. True Health Initiative	http://tinyurl.com/nm2f3cc

The chemical scares industry does a masterful job at perpetuating a myth that was first debunked in the 16th century[30]. This is accomplished by using some rather sophisticated sleight-of-hand:

- Maintaining, or at least suggesting, that a poison is a poison, regardless of the dose or exposure.

- Suggesting that we are constantly bathed in chemicals and that they accumulate in our bodies, where they may cause us harm, such as cancer, birth defects, and autism. This notion is particularly inaccurate, since our bodies are exquisitely designed to break down most chemicals—natural or otherwise—and excrete them. This is the job of your liver.

- Implying that detection of a chemical is proof of harm, or at the least, a cause for alarm. This is based on a false premise. Chemicals that have been in our lives forever can only now be measured, thanks to remarkable advances in the detection power of analytical chemistry. Miniscule amounts of the exact same chemicals that were in our bodies 20 years ago could not be detected then. Does this make us any more or less healthy? No, it does not.

There can no better example of this ploy than the amateurish tactics that Environmental Working Group (EWG) used, presumably to keep us from drinking bottled water. Rather than focus on how silly it is to buy tap water, the environmental burden of manufacturing the plastic bottles, the lack of biodegradation of the plastic, or the fuel required to transport what is essentially a glass of water over long distances, the EWG chose a phony chemical scare tactic. In this case, it employed its ability to measure Tylenol in some samples of bottled water.

30 http://www.toxipedia.org/display/toxipedia/Paracelsus Accessed 12/14/15

In the group's 2008 publication, it tested two brands of bottled water and found acetaminophen (the active component in Tylenol) in both of them[31]. The message, the group claims, seems obvious: Don't drink bottled water since it is chock full of drugs.

The table below may mean nothing to someone who is unfamiliar with concentrations, but to anyone with even a passing knowledge of science, the data in the table are nothing short of ridiculous.

Why? Because since we live in a time when a concentration of parts per billion (ppb) can be measured; bothering to measure anything in parts per trillion borders on insanity. Yet, this did not stop EWG.

To put this table in perspective, it is useful to consider another common chemical—caffeine. Caffeine and acetaminophen have very similar acute toxicity in humans (a lethal dose is in the 10-20 gram range).

The concentration of acetaminophen found in the bottled water, 1.2 ppt, is equivalent to 1.2 micrograms (or, 0.0012 milligrams—8,300 times less than the amount required to equal the weight of one grain of salt) of the drug in a one-liter bottle. Since one tablet of regular strength Tylenol contains 325 mg of acetaminophen, one would need to drink 270 million one-liter bottles of water to get the same amount that is contained in a single pill. This is equivalent to 71 million gallons, or 108 Olympic-size swimming pools.

Perhaps even more absurd, a cup of coffee contains about five ounces, which means that there are seven cups per liter. To ingest a lethal dose of caffeine at a concentration of 51 ppt, one would need to consume about 44,000,000 cups of coffee.

31 http://www.ewg.org/research/bottled-water-quality-investigation/test-results-chemicals-bottled-water 12/22/15

Drugs and drug breakdown products were found in 3 brands

Chemical	Number of Brands	Range of Detection, ppt*	Average of Detected Values, ppt*
Acetaminophen	2	1.1–1.3	1.2
Caffeine	1	51	51
1,7 - Dimethylxanthine (breakdown product of caffeine)	1	10	10

*ppt = parts per trillion (nanograms per liter)

These huge volumes not only point out how outrageously unscientific environmental fundraising groups are, but they also demonstrate how low EWG is willing to stoop to make a phony point that seems to be designed to do nothing except manipulate people for its own purposes.

Despite the obvious absence of risk from consumption of such miniscule quantities of acetaminophen, EWG writes: *"The concentrations in bottled water are below the average therapeutic dosage; however, effects of life-long, constant exposure to this levels of acetaminophen are not known."*

This statement is preposterous. No one will ever perform experiments to determine the life-long effect of virtually immeasurable quantities of acetaminophen in bottled water, because the answer is already known—there is none. It would be an utter waste of time and money to study something is already known to be true according to every legitimate law of science. Not even the EWG wants to conduct this study. Instead of doing science, the group wants to add fuel to their false argument that "any amount of a chemical can hurt you." This is nonsense that Paracelsus, whose conclusion, "the dose makes the poison" provided the foundation of the concept of dose response more than 400 years ago.

4

Intentional misuse of the term "manufactured" and "processed" to imply inferiority

One of the most commonly held misbeliefs among non-scientists is that the way a product is produced speaks in some way to its final composition, properties, or safety profile. After all, who would want their vitamin C coming from a smoke-belching factory in China when it could simply be extracted from delicious, healthy citrus fruit?

While the source of the vitamin C that you consume is irrelevant to your body, we may assume that most consumers are more comfortable with the latter method. If so, 90 percent of people will be disappointed, because of the vitamin C that is consumed in the US, 90 percent of it is manufactured in China[32] from a different source—glucose. By definition, this vitamin C product is both synthetic, and manufactured, but still identical to what is extracted from fruit.

There are a number of methods that have been used to synthesize vitamin C. For many decades, most of it has been manufactured by the Reichstein process, which was invented in the 1930s. The Reichstein process uses glucose as the raw material, and converts it in five synthetic

32 http://www.seattletimes.com/nation-world/china-corners-vitamin-market/ Accessed 12/14/15

steps, all of which use petrochemicals, heavy metals, or acid, to manufacture vitamin C, which is as pure as that from any orange. This material sits in 25 kg cartons ready to be shipped. The factory can manufacture 10 tons of the vitamin per month.

More recently, cost of goods as well as environmental impact have encouraged modifications in the manufacturing process so that the chemical steps have been replaced by fermentation methods—using genetically modified yeast. But, this change may present quite a quandary to the average GNC shopper: Buy something that he almost certainly does not need, and has been made either in a factory from petrochemicals, or in a different factory that uses genetic modification technology.

For a public who shuns anything "artificial"—chemical or biological—this is a thought-provoking mind exercise by any measure.

Conclusion

Chemistry is a complex, often frightening, and sometimes unpopular science. As such, many people instinctively place chemistry in the realm of "something to avoid if at all possible." Indeed, many science students do just that.

But, there are consequences that arise from this mindset; for example, there are individuals, groups, and companies that have profited handsomely by perpetuating myths about chemistry and chemicals. Internet supplement hucksters have thrived by convincing a scientifically naïve public that their "natural" products are not chemicals like prescription drugs and are therefore risk-free. Likewise, the $100 billion organic food industry distinguishes itself from competitors with claims that no synthetic pesticides or herbicides are used when growing its food; the industry fails to note, however, that

The Novachem (Wuhan) Import & Export Company Ltd.
Vitamin C Manufacturing Facility in China

certain naturally derived pesticides are more toxic than their synthetic counterparts, which are used in traditional agriculture.

Certain unscrupulous environmental groups benefit from keeping the American public in the "scientific darkness." Though the Natural Resource Defense Council started less than a decade ahead of the American Council on Science and Health, the NRDC's revenue is 100 times greater; the group needs to constantly create new wars in order to keep that revenue coming in. So, it and others like it constantly launch new fronts in their war on science and health, and one such front is claiming any chemical name is scary.

This new front consists of nothing more than simply manufacturing and spreading fears; the goal is to create concern about trace amounts of harmless chemicals found in our bodies, not because these chemicals weren't present before, but because they simply became detectable thanks to modern technology.

It is also puzzling why environmental groups—the NRDC is perhaps the worst offender—are even interested in trace chemicals in human urine. If trace chemicals represent any issue at all, it is human health—not the environment.

Yet, the group continues to claim that trace chemicals must be harmful even though toxicologists overwhelmingly reject the notion that human exposure to the most miniscule amounts of chemicals is in any way dangerous.

We will leave it to the reader to decide what stake enormous environmental groups like NRDC (annual revenue around $120,000,000) and Greenpeace ($330,000,000) have in throwing out one phony chemical scare after another, and who may be encouraging them to do so.

Finally, it cannot be a surprise that celebrities are cashing in on the anti-chemical movement and playing the "Name Game" as well. Recently,

Vani Hari ("The Food Babe") has been on a one-woman crusade to rid her readers' diets of chemicals, going so far as to spew out utter nonsense:

"When you look at the ingredients, if you can't spell it or pronounce it, you probably shouldn't eat it."[33]

33 http://abcnews.go.com/Health/subway-takes-chemical-sandwich-bread-protest/story?id=22373414 Accessed 12/14/15

Add to this mix Dr. Oz and his notion of "toxic" apple juice (for which he was swatted down by the FDA[34]), Gwyneth Paltrow's random and ridiculous beliefs about health, and, of course, Jenny McCarthy's child-killing campaign against "toxic" vaccines, and it's no wonder that no one knows what to believe any more.

Conventional wisdom indicates that "sticks and stones can break your bones." But, in many cases, names can hurt you too.

34 "Letters from the FDA to the Dr. Oz Show Regarding Apple Juice and Arsenic." http://www.fda.gov/Food/ResourcesForYou/Consumers/ucm271746.htm Accessed 12/14/15

References

Chemicals known to the state to cause cancer or reproductive toxicity, August 2015 http://oehha.ca.gov/prop65/prop65_list/files/P65single082515.pdf Accessed 1/15/16

Cyclooxygenase-3 (COX-3): Filling in the gaps toward a COX continuum?" Warner, T.D., Mitchell, J.A. PNAS(2002), 99(21), 13371-13373.

Dietary supplement health and education act of 1994. Public law 103-417, 103rd congress https://ods.od.nih.gov/About/DSHEA_Wording.aspx#sec1; Accessed 1/21/16

"Ethylene Glycol and Propylene Glycol Toxicity. What is Propylene Glycol?" http://www.atsdr.cdc.gov/csem/csem.asp?csem=12&po=14 Accessed 12/22/15

"Expected New Cancer Cases and Deaths in 2020" http://www.cdc.gov/cancer/dcpc/research/articles/cancer_2020.htm Accessed 12/22/15

How deadly was the poison gas of WW1? BBC News Magazine. http://www.bbc.com/news/magazine-31042472 Accessed 12/22/15

"How Long Do You Have Left to Live at Age 65?" http://politicalcalculations.blogspot.com/2013/01/how-long-do-you-have-left-to-live-at.html - .VqKHp1MwjSc Accessed 12/22/15

http://abcnews.go.com/Health/subway-takes-chemical-sandwich-bread-protest/story?id=22373414 Accessed 12/14/15

http://articles.mercola.com/sites/articles/archive/2015/08/08/toxic-sludge-chemicals.aspx Accessed 12/22/15

http://blog.fooducate.com/2014/01/31/look-at-all-those-chemicals-in-a-banana/ Accessed 12/14/15

http://bodyguardblanket.com/ Accessed 1/15/16

http://www.coinnews.net/2009/05/04/eagle-gold-bullion-coin-sales-strongest-since-1999/ Accessed 1/15/16 http://edition.cnn.com/1999/FOOD/news/12/09/food.prepardness/list.html Accessed 1/24/16

http://www.ewg.org/research/bottled-water-quality-investigation/test-results-chemicals-bottled-water 12/22/15

https://plasticfoodservicefacts.com/Safety-of-Styrene Accessed 1/15/16

http://www.seattletimes.com/nation-world/china-corners-vitamin-market/ Accessed 12/14/15

http://www.sigmaaldrich.com/catalog/product/aldrich/w294004?lang=en&region=US Accessed 11/15/15

http://www.toxipedia.org/display/toxipedia/Paracelsus Accessed 12/14/15

http://www.wholefoodsmarket.com/blog/our-quality-standards-no-artificial-colors-or-flavors Accessed 12/22/15

"If You Think FDA Knows What Chemicals Are In Our Food, Think Again." http://switchboard.nrdc.org/blogs/mmaffini/chemicals_added_to_food.html Accessed 12/22/15

IToxic? http://www.ewg.org/enviroblog/2007/10/itoxic Accessed 12/22/15

"Letters from the FDA to the Dr. Oz Show Regarding Apple Juice and Arsenic." http://www.fda.gov/Food/ResourcesForYou/Consumers/ucm271746.htm Accessed 12/14/15

NaturalNews.com "Healing And True Weight Loss Through Cellular Detoxification." http://blogs.naturalnews.com/healing-and-true-weight-loss-through-cellular-detoxification/ Accessed 12/22/15

"Polystyrene Foam Report" http://www.earthresource.org/campaigns/capp/capp-styrofoam.html Accessed 12/22/15

Q and A Library" Andrew Weil http://www.drweil.com/drw/u/QAA401095/Is-Styrofoam-Safe.html Accessed 12/22/15

"Safety of Polystyrene Foodservice Packaging." Food and Drug Administration: https://plasticfoodservicefacts.com/main/Safety/Safety-of-PS-Foodservice-Products/ Accessed 1/15/16

"School-Associated Violent Death Study." Centers for Disease Control and Prevention, December, 2014 http://www.cdc.gov/violenceprevention/youthviolence/schoolviolence/savd.html Accessed 12/22/15

"Styrofoam Listed As Human Carcinogen" http://www.sustainablebusiness.com/index.cfm/go/news.display/id/25841 Accessed 12/22/15

"The causes of Urban Storm Water Pollution" http://www.nrdc.org/water/pollution/storm/chap2.asp

"The Condition of Education." U.S. Department of Education, National Center for Education Statistics. (2015)

The Day The World Shuts Down, Newsweek, June, 1997 http://www.newsweek.com/day-world-shuts-down-173474 Accessed 1/15/16

"The Food Babe: Enemy of Chemicals. How One Woman Mobilized An Army Against Food Additives, GMOs, And All Else Not "Natural."" Atlantic Magazine 2/11/15 http://www.theatlantic.com/health/archive/2015/02/the-food-babe-enemy-of-chemicals/385301/

"The Thermal Depolymerization Of Polystyrene. Part 2—Formation of "Weak Links"" N. Grassie and W. W. Kerr, *Trans. Faraday Soc.*, 1959, 55,1050-1055.

"The Y2K scare: Causes, Costs and Cures." *Australian Journal Of Public Administration*. Vol. 64, Issue 3, pages 46–55, September 2005.

BOARD OF SCIENTIFIC AND POLICY ADVISORS

BOARD OF SCIENTIFIC AND POLICY ADVISORS (CONTINUED)

BOARD OF SCIENTIFIC AND POLICY ADVISORS (CONTINUED)

BOARD OF SCIENTIFIC AND POLICY ADVISORS (CONTINUED)

The opinions expressed in ACSH publications do not necessarily represent the views of all members of the ACSH Board of Trustees, Founders Circle and Board of Scientific and Policy Advisors, who all serve without compensation.

Made in the USA
Las Vegas, NV
07 November 2026

57650029R00033